Beekeeping for Beginners

Hector A Anderson

Copyright © 2024 **Hector A. Anderson**

All rights reserved. No part of this publication may be reproduced, distributed, or transmitted in any form or by any means, including photocopying, recording, or other electronic or mechanical methods, without the prior written permission of the publisher, except in the case of brief quotations embodied in critical reviews and certain other non-commercial uses permitted by copyright law.

Table of Contents

Introduction 5

Chapter 1 11
- The Life of a Honeybee 11
- Choosing Your Bees 14
- Basic Equipment You Will Need 21

Chapter 2 27
- Choosing the Perfect Location 27
- Setting Up Your Hive 32

Chapter 3 39
- Daily Beekeeping Activities 39
- Feeding Your Bees 45
- Hive Maintenance and Health 49

Chapter 4 56
- Harvesting Honey and Other Products 56
- Other Hive Products 59

Chapter 5 66
- Pest and Disease Management 66
- Hive Management Strategies 72
- Maximizing Honey Production 77

.... **83**

Chapter 6 84
- Inspecting Your Hive 84

Beekeeping Seasons and Annual Hive Care 89

Conclusion 94

Thank you

I am writing to express my sincere gratitude for your interest in my book. Your decision to explore the pages of my work is not only deeply appreciated but also serves as a source of inspiration to me.

I look forward to hearing your thoughts by leaving a honest review on Amazon after exploring the various recipes and other insights in the book. Your feedback is very valuable to me and other authors like me

Introduction

Welcome to "Beekeeping for Beginners," a comprehensive guide designed to accompany you on your exciting journey into the world of beekeeping. Whether you're driven by a desire to produce your own honey, a fascination with the intricate lives of bees, or a commitment to supporting our environment, this book is crafted to provide you with all the knowledge, support, and inspiration you need to become a successful beekeeper.

The Joys of Beekeeping

Beekeeping is more than just a hobby—it's a way to reconnect with nature, contribute to the health of our planet, and engage in a rewarding, lifelong learning experience. There's a profound joy in watching your bees thrive, knowing that you're helping to sustain these vital pollinators. The hum of a busy hive, the sight of bees diligently working on flowers, and the taste of your very own honey are experiences that bring immense satisfaction and fulfillment.

Why Start Beekeeping?

The reasons for starting beekeeping are as diverse as beekeepers themselves. For some, it's the allure of harvesting fresh, natural honey right from their backyard. For others, it's the opportunity to support local agriculture and promote biodiversity. Beekeeping also offers a unique chance to learn and grow, as it combines both scientific inquiry and practical skills. Let's see some reasons why people start keeping bees:

1. Connecting with Nature

In our fast-paced, technology-driven world, beekeeping offers a serene escape back to nature. When you spend time with your bees, you become an integral part of their world. Watching them work, understanding their behavior, and learning to communicate with them fosters a deep appreciation for the natural world. This connection can be profoundly grounding and meditative, offering a peaceful respite from daily stress.

2. Supporting the Environment

Bees play a crucial role in pollinating plants, which is essential for the health of ecosystems and the production of many of the foods we eat. By keeping bees, you contribute to the preservation and support of local biodiversity. Your bees will help pollinate flowers, gardens, and crops in your area, promoting a healthy and balanced environment. Knowing that you're helping sustain these vital pollinators adds a rewarding sense of purpose to beekeeping.

3. Producing Your Own Honey

One of the most tangible rewards of beekeeping is the delicious honey your bees produce. Harvesting honey is a moment of pure joy, the culmination of months of care and attention. This golden, sweet treasure is not only delightful to eat but also makes wonderful gifts for friends and family. Additionally, home-produced honey is often healthier and more flavorful than store-bought varieties, as it's free from additives and processed sugars.

4. Learning and Growth

Beekeeping is a fascinating blend of art and science. As a beekeeper, you'll continually learn new things about bee behavior, hive management, and the complexities of bee biology. This ongoing learning process keeps the hobby engaging and intellectually stimulating. Each season brings new challenges and opportunities to grow your skills, making beekeeping a lifelong journey of discovery.

5. Community and Camaraderie

Beekeeping has a strong, supportive community. By joining local beekeeping clubs or online forums, you can connect with fellow beekeepers, share experiences, and seek advice. These communities are often welcoming and eager to help newcomers. You'll find that beekeepers love to share their passion, and being part of this network can lead to lasting friendships and a sense of belonging.

6. Contributing to Scientific Research

Many beekeepers play a role in scientific research, helping to track bee health and behavior trends. By monitoring your hives and participating in citizen science projects, you can contribute valuable data to the scientific community. This participation can provide a deeper sense of involvement and fulfillment, knowing that you're helping to advance our understanding of these incredible creatures.

7. Financial Benefits

While not typically the primary motivation, beekeeping can offer financial benefits. Selling honey, beeswax, and other hive products can provide a supplementary income. Additionally, some beekeepers offer pollination services to farmers or sell starter hives and bee colonies to new beekeepers. These opportunities can help offset the costs of beekeeping and potentially turn a profit.

8. Teaching Opportunities

Beekeeping provides wonderful opportunities for education. Whether you're teaching your children, engaging in community outreach, or giving talks at local schools, sharing your knowledge about bees can inspire others to appreciate and support these vital insects. Educating others about the importance of bees and beekeeping can foster a greater awareness and promote conservation efforts.

What You Will Learn

This book is structured to guide you step-by-step through the entire beekeeping process, from understanding the basics of bee biology to harvesting honey and maintaining healthy hives. Here's a glimpse of what you can expect to learn:

- **Understanding Bees**: Gain insight into the life of a honeybee, the structure of the hive, and the roles of different bees.

- **Choosing and Setting Up Your Hive**: Learn how to select the right type of bees and equipment, and how to set up your hive for success.

- **Daily Beekeeping Activities**: Discover the essentials of hive inspections, feeding, and maintaining hive health.

- **Harvesting Honey and Other Products**: Master the techniques for extracting honey, beeswax, and other valuable products from your hive.

- **Expanding Your Apiary**: Explore how to grow your beekeeping operation and advance your skills.

- **Community and Education**: Understand the importance of joining beekeeping associations and sharing your knowledge with others.

A Supportive Companion

Beekeeping can be challenging, especially for beginners. This book aims to be more than just an instructional manual; it's a supportive companion that walks with you every step of the way. You'll find practical tips, troubleshooting advice, and words of encouragement to help you navigate the ups and downs of your beekeeping journey.

A Personal Journey

Embarking on beekeeping is also a deeply personal journey. It teaches patience, resilience, and a profound respect for nature. Each moment spent with your bees brings new lessons and rewards. As you flip through the pages of this book, remember that beekeeping is as much about the journey as it is about the destination. Embrace the process, celebrate the small victories, and learn from the challenges.

The Future of Beekeeping

As awareness of the importance of bees grows, more people are turning to beekeeping as a way to make a positive impact. By becoming a beekeeper, you are joining a global movement dedicated to preserving and protecting these vital creatures. The future of beekeeping is bright, filled with opportunities for innovation, conservation, and community building.

Thank you for choosing "Beekeeping for Beginners." I hope this book inspires and equips you to start your beekeeping adventure with confidence and joy. Let's get started on this incredible journey together!

Chapter 1

The Life of a Honeybee

Honeybees are fascinating creatures with complex lives and intricate social structures. Understanding their life cycle and behavior is fundamental to successful beekeeping.

1. The Life Cycle of a Honeybee Honeybees go through four distinct stages in their life cycle: egg, larva, pupa, and adult.

- **Egg**: The queen bee lays each egg in a hexagonal cell within the hive. Depending on the size of the cell, these eggs can develop into worker bees, drones, or new queens.

- **Larva**: After three days, the egg hatches into a larva. Worker bees feed the larva a rich diet of royal jelly for the first few days, followed by a mixture of honey and pollen.

- **Pupa**: The larva spins a cocoon around itself and enters the pupal stage. During this time, it undergoes metamorphosis, transforming into an adult bee.

- **Adult**: Once fully developed, the adult bee emerges from the cell. Worker bees have varied lifespans depending on the season, typically living for a few weeks during busy summer months and longer during the winter.

2. The Daily Life of a Honeybee Adult honeybees have specific roles within the hive, dictated by their age and the needs of the colony.

- **Nurse Bees**: Young worker bees initially serve as nurse bees, tending to the queen and larvae.

- **House Bees**: As they mature, worker bees take on roles such as cleaning the hive, building comb, and processing nectar into honey.

- **Guard Bees**: Some workers become guard bees, defending the hive from intruders.

- **Forager Bees**: In the final stage of their life, worker bees become foragers, venturing out to collect nectar, pollen, water, and propolis.

Roles Within the Hive

The hive is a highly organized society, with each bee playing a specific role to ensure the survival and productivity of the colony.

1. The Queen Bee The queen bee is the heart of the hive. Her primary roles are to lay eggs and produce pheromones that regulate the colony's activities. A healthy queen can lay up to 2,000 eggs per day during peak season. The queen's pheromones also help maintain harmony within the hive, reducing aggression and ensuring cooperation among the workers.

2. Worker Bees Worker bees are female bees with a wide range of responsibilities throughout their lifespan. These roles include:

- **Nursing**: Feeding and caring for the brood.
- **Comb Building**: Producing beeswax and constructing the hexagonal cells of the comb.
- **Cleaning**: Keeping the hive clean and free of debris.
- **Ventilating**: Regulating the hive's temperature by fanning their wings.
- **Foraging**: Collecting nectar, pollen, water, and propolis.

Worker bees are the backbone of the hive, performing the essential tasks that keep the colony functioning smoothly.

3. Drone Bees Drones are male bees whose primary role is to mate with a virgin queen. They are larger than worker bees and lack stingers. During the mating season, drones leave the hive to congregate in drone congregation areas, where they hope to mate with a queen. After mating, drones die, and those that do not mate are expelled from the hive before winter.

The Importance of Bees to the Ecosystem

Honeybees are vital to the health of ecosystems and agriculture worldwide.

1. Pollination Honeybees are among the most effective pollinators. As they forage for nectar and pollen, they transfer pollen from flower to flower, facilitating the fertilization process that leads to the production of fruits, seeds, and nuts. Approximately one-third of the food we eat depends on pollination by bees, including many fruits, vegetables, and nuts.

2. Biodiversity Bees contribute to the diversity and health of plant life. By pollinating a wide variety of plants, they help maintain genetic diversity within plant populations, which is crucial for ecosystem resilience and adaptability.

3. Supporting Wildlife Many animals rely on the fruits, seeds, and plants that bees help pollinate. By supporting plant reproduction, bees indirectly sustain a wide range of wildlife, from insects and birds to mammals.

4. Environmental Indicators Bees serve as important indicators of environmental health. Because they are sensitive to changes in their environment, declines in bee populations can signal broader ecological issues, such as habitat loss, pesticide use, and climate change.

Understanding the life of a honeybee, the roles within the hive, and the broader importance of bees to the ecosystem provides a strong foundation for any aspiring beekeeper. As you delve into beekeeping, this knowledge will not only enhance your appreciation for these remarkable creatures but also guide you in creating a thriving, healthy hive.

Choosing Your Bees

Selecting the right bees is a crucial step in your beekeeping journey. The type of bees you choose can significantly impact your success, as different breeds have varying characteristics and requirements. This chapter will guide you through the different types of bees, how to select the right breed for your needs, and where to purchase your bees.

Different Types of Bees

There are several species and subspecies of honeybees commonly kept by beekeepers. Each type has unique traits that may make it more or less suitable for your specific circumstances.

1. Italian Honeybees (Apis mellifera ligustica)

- **Characteristics**: Italian honeybees are known for their gentle nature and high productivity. They have a yellowish coloration and are relatively easy to manage, making them a popular choice for beginners.

- **Pros**: High honey production, gentle temperament, strong brood rearing.

- **Cons**: Can be prone to robbing other hives and may not handle cold climates as well as other breeds.

2. Carniolan Honeybees (Apis mellifera carnica)

- **Characteristics**: Carniolan bees are dark-colored and known for their calm demeanor and excellent overwintering abilities. They are well-suited to colder climates and tend to build up their populations quickly in the spring.

- **Pros**: Gentle nature, good for cold climates, rapid spring buildup.
- **Cons**: Can swarm more readily if not managed properly.

3. Caucasian Honeybees (Apis mellifera caucasica)

- **Characteristics**: These bees are gray and known for their long tongues, which allow them to access nectar from a variety of flowers. They are gentle and have a strong propensity for propolis production.
- **Pros**: Gentle temperament, efficient foragers, good propolis production.
- **Cons**: High propolis production can make hive management more challenging, slower spring buildup.

4. Buckfast Honeybees

- **Characteristics**: Buckfast bees are a hybrid breed developed by Brother Adam of Buckfast Abbey. They are known for their resilience to disease, high honey production, and gentle nature.
- **Pros**: Disease resistance, high productivity, gentle nature.
- **Cons**: Can be more expensive and harder to source.

5. Russian Honeybees

- **Characteristics**: Russian bees are known for their resistance to varroa mites and their ability to thrive in cold climates. They have a more variable temperament and are good at managing their own population.

- **Pros**: Varroa mite resistance, good overwintering, efficient brood management.

- **Cons**: Variable temperament, can be more defensive.

6. Africanized Honeybees (Apis mellifera scutellata hybrid)

- **Characteristics**: Also known as "killer bees," Africanized honeybees are known for their aggressive behavior and high adaptability. They are not recommended for beginners due to their defensive nature.

- **Pros**: High adaptability and productivity.

- **Cons**: Extremely aggressive, not suitable for beginners or populated areas.

How to Select the Right Breed for You

Choosing the right breed of bees depends on various factors, including your location, experience level, and specific goals as a beekeeper. Here are some considerations to help you make the best choice:

1. Climate

- **Cold Climates**: Carniolan and Russian bees are excellent choices due to their strong overwintering abilities.

- **Warm Climates**: Italian and Africanized bees can thrive in warmer environments, though caution is advised with Africanized bees due to their aggressive nature.

2. Experience Level

- **Beginners**: Italian and Buckfast bees are highly recommended for their gentle nature and ease of management.

- **Intermediate/Advanced**: Carniolan, Caucasian, and Russian bees offer unique advantages but may require more attentive management.

3. Honey Production If high honey production is your primary goal, consider Italian, Buckfast, or Russian bees, known for their productivity.

4. Disease Resistance For areas with high disease pressure, Russian and Buckfast bees are excellent choices due to their resilience.

5. Swarming Tendencies If you're concerned about managing swarming, Carniolan bees might require more intervention, while Italian bees tend to be less swarm-prone.

6. Propolis Production If you prefer less propolis in your hives, avoid Caucasian bees, as they are known for their high propolis production.

Where to Buy Bees

Once you've decided on the type of bees that best suit your needs, the next step is to purchase them. Here are some options for acquiring your bees:

1. Local Beekeepers

- **Pros**: Buying from local beekeepers can ensure that the bees are already adapted to your local climate and conditions. It also supports your local beekeeping community.

- **Cons**: Availability may be limited, and there may be a waiting list.

2. Beekeeping Associations

- **Pros**: Many beekeeping associations have programs to help new beekeepers get started, including selling bees or providing them at a reduced cost.

- **Cons**: Availability can be seasonal and dependent on the association's resources.

3. Bee Suppliers and Apiaries

- **Pros**: Commercial bee suppliers offer a variety of bee breeds and can often ship bees directly to you. This provides flexibility and access to different types of bees.

- **Cons**: Shipping can be stressful for the bees, and there's a risk of introducing bees that are not well adapted to your local environment.

4. Nucleus Colonies (Nucs)

- **Pros**: Nucs are small, established colonies with a queen, workers, and brood. They are often easier to manage initially than packages of bees.

- **Cons**: Nucs can be more expensive than packages of bees and may be less readily available.

5. Package Bees

- **Pros**: Packages typically include a queen and a few pounds of worker bees. They are widely available and often more affordable than nucs.

- **Cons**: It can take time for package bees to establish a new colony, and there may be a higher initial mortality rate.

6. Swarm Traps

- **Pros**: Capturing a wild swarm is a cost-effective way to start a hive and can be very rewarding.

- **Cons**: Swarm trapping can be unpredictable, and there is a risk of capturing bees with undesirable traits or diseases.

Choosing the right bees and sourcing them effectively is the foundation of a successful beekeeping venture. With the right breed suited to your specific needs and conditions, you'll be well on your way to creating a thriving, productive hive.

Basic Equipment You Will Need

Starting your beekeeping journey requires some essential equipment to ensure you and your bees thrive. This chapter covers the fundamental items you'll need, including hives and frames, protective gear, and various tools and accessories.

Hives and Frames

The hive is the heart of your beekeeping operation. It provides shelter, a place for bees to build their comb, and a structure for you to manage and harvest honey.

1. Hives There are several types of hives, but the most common for beginners are the Langstroth, Top-Bar, and Warre hives.

- Langstroth Hive: The most widely used hive, it consists of stacked rectangular boxes with removable frames. It's highly efficient for honey production and hive inspections.

Pros: Standardized equipment, easy to expand, efficient honey production.

Cons: Heavier and more complex to manage.

- Top-Bar Hive: A horizontal hive with bars across the top for bees to build their comb. It's more natural for bees and easier to manage without heavy lifting.

Pros: Easy to manage, minimal lifting, natural comb building.

Cons: Less honey production, non-standard equipment.

- Warre Hive: A vertical hive similar to the Langstroth but with smaller, square boxes. It's designed to mimic a tree cavity and promote natural beekeeping practices.

Pros: Mimics natural bee habitats, minimal intervention.

Cons: Less honey production, more difficult to inspect.

2. Frames Frames hold the honeycomb and fit into the hive boxes. For Langstroth hives, they come in different sizes: deep, medium, and shallow.

- **Deep Frames**: Used in brood boxes where the queen lays eggs.
- **Medium Frames**: Versatile and can be used for both brood and honey.
- **Shallow Frames**: Primarily used for honey supers, making them easier to handle when full.

Frames can come with or without foundation, a pre-formed sheet that guides bees in building comb. Foundation can be wax or plastic.

3. Foundation Foundation provides a base for bees to build their comb. It can be made of beeswax or plastic.

- **Wax Foundation**: Natural and preferred by bees but can be fragile.

- **Plastic Foundation**: Durable and easy to use but less natural.

Protective Gear

Safety and comfort are essential when working with bees. Proper protective gear will help you feel confident and safe.

1. Bee Suit A full bee suit provides the best protection, covering your entire body.

- **Ventilated Bee Suits**: Made with layers of breathable material, they keep you cool while providing sting protection.
- **Standard Bee Suits**: Single-layer suits that offer good protection but can be hot in warm weather.

2. Gloves Bee gloves protect your hands and wrists.

- **Leather Gloves**: Durable and offer good protection, but can reduce dexterity.
- **Nitrile Gloves**: Provide better dexterity and are still effective against stings, but can be less durable.

3. Veil A bee veil protects your face and neck, the most vulnerable areas.

- **Hooded Veil**: Attached to a suit, offering comprehensive protection.
- **Hat and Veil**: Separate pieces that can be used with a jacket.

4. Boots High boots or boot covers protect your ankles and lower legs.

- **Bee Boots**: Designed to prevent bees from entering, usually worn over your regular footwear.

- **Work Boots**: Sturdy and practical, providing good protection when combined with bee suits that cover your ankles.

Tools and Accessories

Having the right tools and accessories makes hive management easier and more efficient.

1. Hive Tool A hive tool is indispensable for beekeeping tasks such as prying apart hive boxes, scraping off propolis, and removing frames.

- **Standard Hive Tool**: A flat, metal tool with a hooked end and a flat end.

- **J-Hook Hive Tool**: Features a hook that makes it easier to lift frames.

2. Smoker A smoker calms bees by masking alarm pheromones and encouraging them to retreat into the hive.

- **Stainless Steel Smoker**: Durable and commonly used, with a heat guard to protect your hands.

- **Fuel**: Common fuels include pine needles, burlap, and smoker pellets.

3. Bee Brush A soft-bristled brush for gently removing bees from frames without harming them.

4. Frame Grip A tool that helps you lift and handle frames with ease, particularly useful when frames are heavy with honey.

5. Queen Catcher and Marking Tools

- **Queen Catcher**: A small device for safely capturing the queen during inspections.
- **Queen Marking Pen**: Used to mark the queen for easy identification.

6. Feeder Feeders provide supplementary nutrition to bees when natural forage is scarce.

- **Boardman Feeder**: Fits at the entrance of the hive.
- **Top Feeder**: Sits on top of the frames inside the hive.
- **Frame Feeder**: Replaces one of the frames inside the hive.

7. Bee Escape A device that allows bees to exit a honey super but not re-enter, making honey harvesting easier.

8. Honey Extractor Used to extract honey from the frames.

- **Manual Extractor**: Requires hand-cranking, suitable for small-scale beekeeping.
- **Electric Extractor**: Easier and faster, ideal for larger operations.

Equipping yourself with the right hives, protective gear, and tools sets the foundation for successful and enjoyable beekeeping. As you gain experience, you may find additional tools and equipment that suit your specific needs, but starting with these basics will ensure you're well-prepared to manage and care for your bees.

Chapter 2

Choosing the Perfect Location

Selecting the right location for your beehives is critical to the health and productivity of your bees. This chapter will guide you through the essential aspects of choosing the perfect location, including space requirements, environmental factors, and safety considerations.

Space Requirements

Proper spacing is vital for both the health of your bees and the ease of hive management.

1. Adequate Space Around Hives

Hive Placement: Place hives at least two to three feet apart to allow easy access for inspections and maintenance.

Work Area: Ensure there's enough space around each hive for you to work comfortably. A clear area of at least six feet around the hive is ideal.

2. Hive Stand or Platform

Elevate Hives: Use a hive stand or cinder blocks to elevate your hives off the ground. This protects the hives from moisture, pests, and improves ventilation.

Level Ground: Make sure the ground is level to prevent the hives from tipping over and to ensure proper drainage.

3. Access for Maintenance

Pathways: Create clear pathways to your hives for easy access. These paths should be wide enough for you to carry equipment and harvested honey.

Vehicle Access: If possible, choose a location that allows vehicle access for transporting heavy supplies and harvested honey.

Environmental Factors

Environmental conditions play a significant role in the well-being of your bees. Consider the following factors when choosing a location:

1. Sunlight

Morning Sun: Place your hives where they will receive morning sunlight. This helps warm the hive early in the day, encouraging bees to start foraging sooner.

Partial Shade: In hot climates, some afternoon shade can prevent the hive from overheating. Use natural shade from trees or artificial shade structures if necessary.

2. Wind Protection

Windbreaks: Protect your hives from strong winds by placing them near natural windbreaks such as trees, shrubs, or fences. This helps maintain a stable hive temperature and reduces stress on the bees.

Orientation: Position the hive entrance away from prevailing winds to make it easier for bees to come and go.

3. Water Source

Proximity to Water: Bees need a reliable source of water for drinking, cooling the hive, and diluting honey. Place your hives near a clean, accessible water source such as a pond, stream, or birdbath.

Artificial Water Source: If no natural water source is available, provide a shallow dish of water with stones or floating corks for bees to land on.

4. Forage Availability

Diverse Forage: Ensure your bees have access to a variety of flowering plants throughout the growing season. Bees forage within a three-mile radius of their hive, so a diverse landscape with continuous blooms is ideal.

Avoiding Pesticides: Choose a location away from areas where pesticides are heavily used, such as commercial agricultural fields or golf courses, to prevent harm to your bees.

5. Drainage

Avoid Low Spots: Place your hives in a well-drained area to prevent water from pooling around the hive, which can cause damp conditions and increase the risk of diseases.

Slope: A slight slope can help with drainage, ensuring the area around the hive remains dry.

Safety Considerations

Safety for both you and your bees is paramount when choosing a hive location.

1. Distance from Residences and Public Areas

Buffer Zone: Place your hives away from homes, schools, playgrounds, and public pathways to minimize the risk of bee encounters with people.

High Traffic Areas: Avoid placing hives near high-traffic areas to reduce the likelihood of disturbing the bees and causing defensive behavior.

2. Barriers and Fencing

Fencing: Erect a fence or hedge around your apiary to direct the bees' flight path upward and away from nearby areas. A six-foot barrier can effectively guide bees to fly above head height.

Visual Barriers: Use visual barriers such as shrubs or lattice panels to create a buffer zone and reduce direct line-of-sight to the hives.

3. Legal and Zoning Requirements

Local Regulations: Check local ordinances and zoning laws regarding beekeeping. Some areas have specific requirements for hive placement, number of hives, and registration.

Neighbors: Inform your neighbors about your beekeeping plans.

Address any concerns they might have and educate them about the benefits of bees.

4. Predators and Pests

Predator Protection: Place hives where they are less likely to be disturbed by predators such as skunks, bears, or raccoons. In bear country, use electric fencing to protect your hives.

Ant Control: Avoid placing hives directly on the ground, as this can make them more susceptible to ant infestations. Use hive stands and apply barriers such as diatomaceous earth around the hive stand legs.

5. Personal Safety

Bee Allergies: Ensure you and anyone regularly visiting the apiary are aware of the potential for bee stings and have access to medical care if necessary. Those with severe allergies should carry an epinephrine auto-injector.

Protective Gear: Always wear appropriate protective gear when working with your hives to minimize the risk of stings.

Choosing the perfect location for your beehives involves careful consideration of space, environmental factors, and safety. By taking the time to select an optimal site, you'll create a supportive environment where your bees can thrive and produce honey, ensuring a successful and rewarding beekeeping experience.

Setting Up Your Hive

Setting up your hive is an exciting step in your beekeeping journey. Proper assembly, installation, and initial care are crucial for establishing a healthy and productive colony. This chapter will guide you through assembling the hive, installing the bees, and providing initial feeding and care.

Assembling the Hive

Before your bees arrive, it's essential to have your hive assembled and ready. Here's a step-by-step guide to help you set up your hive correctly.

1. Hive Components

- **Bottom Board**: The base of the hive, which can be solid or screened for ventilation.

- **Hive Boxes (Supers)**: These include deep boxes for the brood chamber and medium or shallow boxes for honey storage.

- **Frames**: Removable frames that hold the foundation for the bees to build comb.

- **Inner Cover**: Placed on top of the uppermost hive box to provide insulation and ventilation.

- **Outer Cover**: The roof of the hive, protecting it from weather conditions.

2. Assembly Steps

- **Step 1: Bottom Board Placement**: Position the bottom board on your hive stand or platform. Ensure it is level and stable.

- **Step 2: Hive Boxes**: Place the first deep box on the bottom board. This will be your brood chamber where the queen lays eggs.

- **Step 3: Frames**: Insert the frames into the hive box. Ensure they are evenly spaced and fit snugly.

- **Step 4: Additional Boxes**: As your colony grows, you'll add more hive boxes (supers) above the brood chamber. Initially, one brood box is sufficient.

- **Step 5: Inner and Outer Covers**: Place the inner cover on top of the uppermost box, followed by the outer cover.

3. Securing the Hive

- **Ratchet Straps**: Use ratchet straps to secure the hive components together, especially if you live in a windy area or are transporting the hive.

- **Entrance Reducer**: Install an entrance reducer to control the size of the hive entrance, helping the bees defend against pests and regulate temperature.

Installing the Bees

Once your hive is assembled, it's time to install your bees.

Whether you're starting with a package of bees or a nucleus colony (nuc), proper installation is key to helping your bees settle in.

1. Installing a Package of Bees

- **Preparation**: Before installing the bees, ensure your hive is set up and ready. Have a spray bottle with sugar water (1:1 ratio) and a hive tool on hand.

- **Step 1: Spray the Bees**: Lightly spray the bees in the package with sugar water to calm them and make them less likely to fly.

- **Step 2: Remove the Queen Cage**: Carefully remove the queen cage from the package. Inspect the queen to ensure she is alive and healthy.

- **Step 3: Release the Queen**: There are two methods to release the queen: direct release or delayed release. For beginners, the delayed release method is recommended.

Delayed Release: Remove the cork from the candy end of the queen cage. Place the queen cage between two frames in the brood box, ensuring the candy end is accessible to the worker bees.

- **Step 4: Install the Bees**: Gently shake the package to dislodge the bees into the hive. Pour them over the frames in the brood box.

- **Step 5: Close the Hive**: Replace the inner cover and outer cover. Leave the bees undisturbed for a few days to acclimate to their new home.

2. Installing a Nucleus Colony (Nuc)

- **Preparation**: Ensure your hive is ready, and have a hive tool and protective gear on hand.

- **Step 1: Transfer Frames**: Open the nuc box and carefully transfer the frames, one by one, into your hive box. Maintain the same order and orientation of the frames.

- **Step 2: Fill the Hive Box**: Add additional frames to fill the hive box if necessary.

- **Step 3: Close the Hive**: Place the inner cover and outer cover on the hive. Allow the bees to settle in without disturbance for a few days.

Initial Feeding and Care

Providing proper nutrition and care during the initial stages is crucial for the colony's development and health.

1. Feeding the Bees

- **Sugar Syrup**: Prepare a 1:1 sugar syrup (one part sugar to one part water) to feed your bees. Use a hive top feeder, frame feeder, or entrance feeder to provide the syrup.

- **Frequency**: Feed the bees until they have established comb and are actively foraging. This usually takes a few weeks.

2. Monitoring and Inspections

- **Initial Inspection**: Conduct a quick inspection a few days after installing the bees to ensure the queen is released (if using the delayed release method) and the bees are settling in.

- **Regular Inspections**: Perform weekly inspections to monitor the colony's progress, check for signs of disease or pests, and ensure the queen is laying eggs.

3. Temperature and Ventilation

- **Temperature Control**: Ensure the hive is well-ventilated, especially in hot weather. Screened bottom boards and vented inner covers can help regulate temperature.

- **Winter Preparations**: If you install bees in early spring, prepare for the winter by ensuring the hive is well-insulated and has enough stored honey for the bees to survive.

4. Pest and Disease Management

- **Pest Control**: Monitor for common pests like varroa mites and small hive beetles. Use appropriate treatments and management practices to keep these pests under control.

- **Disease Prevention**: Keep an eye out for signs of diseases such as American foulbrood or nosema. Maintain good hive hygiene and practice regular inspections to catch issues early.

Setting up your hive correctly and providing the necessary initial care will give your bees the best start in their new home. With a solid foundation and attentive management, your colony will thrive and reward you with the joys of beekeeping.

Chapter 3

Daily Beekeeping Activities

Daily beekeeping activities are essential for maintaining a healthy and productive hive. Regular inspections help you monitor the health of your colony, identify and address issues early, and ensure the hive is functioning properly. This chapter will guide you through how to conduct a hive inspection, what to look for, and common problems with their solutions.

Inspecting the Hive

Regular hive inspections are vital for understanding your bees' needs and maintaining a thriving colony. Here's how to conduct a thorough and effective hive inspection.

1. How to Conduct a Hive Inspection

Proper preparation and technique are crucial for a successful hive inspection.

Step 1: Gather Your Tools and Gear

- **Protective Gear**: Wear your bee suit, gloves, and veil to protect yourself from stings.

- **Hive Tool**: Essential for prying apart hive components and handling frames.

- **Smoker**: Calms the bees and reduces defensive behavior. Use natural fuel like pine needles or burlap.

- **Notebook or App**: Record your observations for future reference.

Step 2: Prepare the Smoker

- Light the smoker and ensure it's producing cool, steady smoke. Have it ready before you approach the hive.

Step 3: Approach the Hive Calmly

- Move slowly and avoid sudden movements. Approach the hive from the side or back to avoid blocking the entrance.

Step 4: Use the Smoker

- Puff a few gentle clouds of smoke at the hive entrance and under the lid. Wait a minute to allow the smoke to calm the bees.

Step 5: Open the Hive

- Remove the outer cover and give a few more puffs of smoke under the inner cover. Gently pry the inner cover off with your hive tool.

Step 6: Inspect the Frames

- Start with the outer frames and work towards the center. Carefully lift each frame, hold it over the hive, and examine both sides.
- Look for signs of brood (eggs, larvae, pupae), honey,

pollen, and the queen. Handle the frames gently to avoid crushing bees or damaging comb.

Step 7: Record Observations

- Note the presence and condition of the queen, brood pattern, honey stores, and any signs of pests or diseases. Recording this information helps track the hive's progress and identify trends.

Step 8: Close the Hive

- Once the inspection is complete, reassemble the hive in reverse order. Ensure all parts fit snugly and the entrance is clear.

Step 9: Clean and Store Tools

- Clean your hive tool and smoker, and store your gear in a dry, secure place.

2. What to Look For

During a hive inspection, look for specific indicators of hive health and productivity.

Queen Presence and Activity

- **Queen Sightings**: Look for the queen herself or evidence of her presence, such as eggs and young larvae.

- **Brood Pattern**: A healthy queen lays eggs in a consistent, solid pattern. Spotty brood can indicate queen issues or disease.

Brood Health

- **Eggs and Larvae**: Check for eggs (small, white, rice-like objects) and larvae (white, curled in the cells).

- **Pupae**: Capped brood cells should have a uniform appearance. Sunken or perforated caps can indicate disease.

Honey and Pollen Stores

- **Honey**: Look for capped honey in the upper frames. Ensure there's enough for the bees' needs, especially before winter.

- **Pollen**: Pollen stores are usually found near the brood and are essential for feeding larvae.

Hive Condition

- **Comb Structure**: Healthy, clean comb is vital. Look for any irregularities, excessive burr comb, or damaged frames.

- **Hive Cleanliness**: Bees are meticulous about hive cleanliness. Excessive debris or mold can indicate issues.

Signs of Pests and Diseases

- **Varroa Mites**: Check for mites on bees and in drone brood cells. Look for signs like deformed wings.

- **Small Hive Beetles**: Look for adult beetles and larvae in the hive.

- **Wax Moths**: Signs include webbing, larvae, and damage to the comb.

- **Diseases**: Symptoms of diseases like American Foulbrood (AFB) or Nosema should be identified early. AFB causes discolored, sunken brood caps, and a foul smell.

3. Common Problems and Solutions

Identifying and addressing common problems early can prevent them from escalating.

Problem: Varroa Mites

- **Symptoms**: Mites on bees, deformed wings, reduced brood.

- **Solutions**: Regular monitoring, using mite treatments (chemical or natural), and maintaining strong colonies.

Problem: Small Hive Beetles

- **Symptoms**: Beetles in the hive, larvae in comb, slimy frames.

- **Solutions**: Beetle traps, maintaining strong colonies, and reducing hive moisture.

Problem: Wax Moths

- **Symptoms**: Webbing, larvae in comb, damaged frames.
- **Solutions**: Regular inspections, maintaining strong colonies, and storing spare equipment properly.

Problem: American Foulbrood (AFB)

- **Symptoms**: Sunken brood caps, foul smell, sticky brown larval remains.
- **Solutions**: Remove and burn infected frames, treat with antibiotics if legal and recommended, and practice good hygiene.

Problem: Nosema

- **Symptoms**: Diarrhea on hive entrance, reduced population, slow build-up.
- **Solutions**: Maintain dry, clean hives, use fumagillin if legal and recommended, and replace old comb regularly.

Problem: Swarming

- **Symptoms**: Queen cells, overcrowded hive, reduced foraging.
- **Solutions**: Regular hive inspections, splitting strong colonies, providing additional space.

Conducting regular hive inspections and knowing what to look for ensures you maintain a healthy and productive colony. By addressing issues promptly and effectively, you'll create a stable environment for your bees to thrive, contributing to a successful and rewarding beekeeping experience.

Feeding Your Bees

Feeding your bees is essential for ensuring they have an adequate food supply, especially during times of scarcity or when starting a new colony. This chapter will guide you through when and what to feed your bees, as well as the differences between natural and supplemental feeding.

When and What to Feed

Knowing when and what to feed your bees is crucial for maintaining their health and productivity.

1. Times to Feed Your Bees

Early Spring: As winter ends and resources are scarce, feeding your bees in early spring helps stimulate brood rearing and colony growth.

Late Summer/Fall: In late summer and fall, when natural forage may be limited, providing supplemental feeding helps bees build up their winter stores.

Winter: Depending on your climate and the amount of honey stores, you may need to feed your bees throughout the winter to ensure they have enough food to survive until spring.

2. What to Feed Your Bees

Sugar Syrup: A mixture of sugar and water is the most common feed for bees. The ratio of sugar to water varies depending on the purpose:

- **Spring Feeding**: Use a 1:1 ratio (1 part sugar to 1 part water) to stimulate brood production and colony growth.
- **Fall Feeding**: Use a 2:1 ratio (2 parts sugar to 1 part water) to provide a more concentrated energy source for winter storage.

Fondant or Candy: Solid sugar sources like fondant or candy can be placed directly on the frames for bees to access during times of low forage availability.

Pollen Patties: Pollen patties provide essential protein for brood rearing, especially in early spring when natural pollen sources may be scarce.

Natural vs. Supplemental Feeding

Understanding the differences between natural and supplemental feeding helps you make informed decisions about providing food for your bees.

1. Natural Feeding

Pros:

- Mimics natural foraging behavior.

- Bees collect a diverse range of nutrients from natural sources.

- Supports overall colony health and genetic diversity.

Cons:

- Relies on the availability of natural forage, which can be unpredictable.

- Limited control over the quality and quantity of food sources.

- May not provide sufficient nutrition during periods of scarcity.

2. Supplemental Feeding

Pros:

- Provides a reliable food source during times of scarcity or for new colonies.

- Allows beekeepers to control the quality and quantity of food provided.

- Helps ensure bees have enough resources to thrive and survive winter.

Cons:

- Can disrupt natural foraging behavior and genetic diversity.

- Overfeeding can lead to issues such as sugar syrup fermentation or excess moisture in the hive.
- Requires additional time and resources for preparation and management.

Tips for Effective Feeding

1. Timing: Feed your bees when natural forage is scarce or during periods of colony growth, such as early spring or fall.

2. Observation: Monitor hive activity and food stores regularly to determine when feeding is necessary.

3. Quality: Use high-quality sugar and water for sugar syrup, and ensure pollen patties are fresh and free from contaminants.

4. Placement: Place feeders or supplemental food close to the brood nest to encourage bees to access it readily.

5. Monitoring: Keep an eye on hive weight and food consumption to ensure bees have enough stores for their needs.

6. Avoidance: Minimize feeding during periods of nectar flow to avoid disrupting natural foraging behavior and potentially diluting honey quality.

Feeding your bees is a crucial aspect of beekeeping, ensuring they have the resources they need to thrive and remain healthy. By understanding when and what to feed your bees and the differences between natural and supplemental feeding, you can effectively support your colonies throughout the year, contributing to a successful and rewarding beekeeping experience.

Hive Maintenance and Health

Maintaining the health of your hive is essential for the well-being of your bees and the success of your beekeeping operation. This chapter covers seasonal maintenance tasks, recognizing and treating diseases, and managing pests and predators to ensure your hive remains strong and productive.

Seasonal Maintenance

Proper seasonal maintenance keeps your hive in optimal condition throughout the year, supporting the health and productivity of your bees.

Spring

- **Hive Inspection**: Conduct thorough inspections to assess hive strength, brood production, and food stores. Stimulate brood rearing by feeding sugar syrup if necessary.

- **Swarm Prevention**: Manage hive overcrowding by splitting strong colonies or adding supers. Remove queen cells to prevent swarming.

- **Varroa Monitoring**: Begin regular varroa mite monitoring and treatment if mite levels exceed recommended thresholds.

Summer

- **Supers and Honey Harvest**: Add honey supers as needed and harvest honey when frames are capped and ready. Ensure ample ventilation to prevent overheating.

- **Water Source**: Ensure bees have access to a clean water source to prevent dehydration and reduce swarming behavior.

- **Pest Control**: Monitor for pests such as small hive beetles and wax moths. Use traps or treatments as necessary.

Fall

- **Winter Preparation**: Inspect and assess hive strength and food stores. Ensure bees have enough honey for winter survival. Consider supplemental feeding if necessary.

- **Varroa Treatment**: Treat for varroa mites using appropriate methods before winter bees emerge.

- **Hive Insulation**: Provide insulation such as hive wraps or foam boards to help bees conserve heat during winter.

Winter

- **Hive Monitoring**: Check hive weight and food stores periodically. Ensure entrances remain clear of snow and ice.

- **Ventilation**: Maintain adequate ventilation to prevent moisture buildup and condensation inside the hive.

- **Emergency Feeding**: If food stores are low, provide emergency feeding with fondant or candy boards to prevent starvation.

Recognizing and Treating Diseases

Early detection and prompt treatment of diseases are critical for preserving hive health and preventing the spread of pathogens.

American Foulbrood (AFB)

- **Symptoms**: Sunken, perforated brood cappings with a foul odor. Larvae may appear brown and ropey.

- **Treatment**: Burn affected frames and colonies. Antibiotic treatment may be used, but regulations vary by location.

Nosema

- **Symptoms**: Diarrhea on hive entrance, reduced colony strength, slow build-up in spring.

- **Treatment**: Fumagillin treatment, maintain good hive hygiene, and replace old comb regularly.

Varroa Mites

- **Symptoms**: Deformed wings, weakened bees, and increased winter mortality.
- **Treatment**: Integrated pest management (IPM) strategies, including chemical treatments, organic acids, and mechanical methods such as drone brood removal.

European Foulbrood (EFB)

- **Symptoms**: Discolored, melted-looking brood with a foul odor. Larvae may appear brown and slimy.
- **Treatment**: Strengthen colonies, requeen if necessary, and improve hive hygiene. Antibiotic treatment may be used in severe cases.

Chalkbrood

- **Symptoms**: Hard, chalk-like mummies in brood cells. Reduced brood production and weakened colonies.
- **Treatment**: Improve hive ventilation and hygiene, requeen if necessary, and remove affected comb.

Managing Pests and Predators

Effective pest and predator management helps protect your hive and ensure the well-being of your bees.

Varroa Mites

- **Monitoring**: Regularly monitor mite levels using sticky boards or alcohol washes.

- **Treatment**: Implement an integrated pest management (IPM) approach, including chemical treatments, organic acids, and mechanical methods.

Small Hive Beetles

- **Traps**: Use beetle traps filled with oil to catch and
- remove adult beetles.
- **Hive Hygiene**: Maintain good hive hygiene to reduce beetle populations. Keep the hive strong and well-ventilated.

Wax Moths

- **Prevention**: Maintain strong colonies with healthy populations. Keep hive components clean and free from debris.
- **Traps**: Use pheromone traps to attract and capture adult wax moths.

Bears and Other Predators

- **Fencing**: Install electric fencing around the hive to deter bears and other large predators.
- **Secure Hives**: Ensure hive components are securely fastened to prevent them from being easily knocked over or accessed by predators.

Regular hive maintenance, disease recognition, and pest management are essential aspects of responsible beekeeping.

By following seasonal maintenance tasks, promptly treating diseases, and implementing effective pest management strategies, you can ensure the health and longevity of your hive, ultimately leading to a successful and rewarding beekeeping experience.

Chapter 4

Harvesting Honey and Other Products

Harvesting honey is one of the most rewarding aspects of beekeeping, providing you with delicious, natural sweetener straight from your own hive. This chapter will cover when to harvest honey, how to extract and process it, and tips for storing and using your honey effectively.

Honey Harvesting

When to Harvest Honey

Timing your honey harvest ensures that you collect mature, capped honey with optimal flavor and moisture content.

1. Assessing Honey Ripeness

Capped Cells: Wait until at least 80% of the cells in the honeycomb are capped with beeswax. This indicates that the honey has been properly cured and is ready for harvest.

Moisture Content: Use a refractometer to measure the moisture content of the honey. Ideally, honey should have a moisture content of around 18% or lower for long-term storage.

2. Environmental Factors

Weather: Choose a dry, sunny day for harvesting to minimize moisture absorption and prevent honey from fermenting.

Forage Availability: Consider the availability of nectar sources in your area. Harvest honey after the main nectar flow when bees have finished collecting and capping honey.

How to Extract and Process Honey

1. Preparation

Remove Supers: Remove honey supers from the hive, ensuring that bees are not present.

Uncap Cells: Use a hot knife or uncapping fork to remove the thin beeswax capping from the honeycomb frames.

2. Extraction

Extractor: Place uncapped frames in a honey extractor—a centrifuge that spins the frames, causing honey to fly out of the cells.

Strain: Filter the extracted honey through a fine mesh or cheesecloth to remove any remaining wax or debris.

3. Bottling

Storage Containers: Transfer the strained honey into clean, dry storage containers, such as glass jars or plastic bottles.

Labeling: Label each container with the harvest date and honey variety if applicable.

Storing and Using Your Honey

1. Storage

Cool, Dark Place: Store honey in a cool, dark place away from direct sunlight and heat sources to prevent crystallization and flavor degradation.

Airtight Containers: Seal containers tightly to prevent moisture absorption and maintain honey quality.

2. Culinary Uses

Sweetener: Use honey as a natural sweetener in teas, baked goods, dressings, and marinades.

Flavor Enhancer: Experiment with different honey varieties to add unique flavors to dishes and beverages.

3. Health and Wellness

Natural Remedy: Honey has antimicrobial properties and is often used as a remedy for sore throats and coughs.

Skincare: Apply honey topically as a moisturizing face mask or as a treatment for minor cuts and burns.

4. Gift Giving

Homemade Gifts: Package honey in decorative jars or bottles to give as thoughtful gifts for family and friends.

Personalization: Create custom labels or tags to add a personal touch to your honey gifts.

Harvesting honey is a rewarding culmination of your efforts as a beekeeper. By timing your harvest correctly, extracting and processing honey with care, and storing it properly, you can enjoy the sweet fruits of your labor throughout the year. Whether you use honey in culinary creations, as a natural remedy, or as a heartfelt gift, the delicious and versatile nature of honey adds joy to every aspect of beekeeping.

Other Hive Products

In addition to honey, bees provide a variety of other valuable products, each with unique properties and uses. This chapter will explore beeswax, propolis, and royal jelly, including how to harvest and utilize these products.

Beeswax

1. What is Beeswax?

Beeswax is a natural wax produced by honeybees from glands on the underside of their abdomen. Bees use it to construct the hexagonal cells of the honeycomb, which house brood and store honey and pollen.

2. Harvesting Beeswax

Honeycomb Cappings: The most common source of beeswax is the cappings that bees seal over honey cells. These are removed during the honey extraction process.

- **Step 1: Collect Cappings**: As you uncap the honeycomb, collect the beeswax cappings in a separate container.

- **Step 2: Clean the Wax**: Rinse the cappings to remove any residual honey. Melt the wax in a double boiler or solar wax melter, then strain it through cheesecloth to remove impurities.

Old Comb: Beeswax can also be harvested from old brood comb, which should be replaced periodically to maintain hive health.

- **Step 1: Remove Old Comb**: Cut out old, darkened comb from frames.
- **Step 2: Clean and Melt**: Similar to cappings, rinse, melt, and strain the wax to purify it.

3. Uses of Beeswax

Candle Making: Beeswax candles burn cleanly and slowly, producing a pleasant natural scent.

- **Process**: Melt purified beeswax and pour it into molds with wicks, or use it to dip taper candles.

Cosmetics and Skincare: Beeswax is a common ingredient in lip balms, lotions, and creams due to its moisturizing properties.

- **Recipes**: Combine beeswax with oils, butters, and essential oils to create homemade skincare products.

Food Wraps: Beeswax-coated fabric wraps are a sustainable alternative to plastic wrap.

- **DIY**: Melt beeswax and brush it onto cotton fabric. Once it cools, the wrap can be used to cover food containers or wrap produce.

Wood Polish: Beeswax can be used to create natural wood polish.

- **Recipe**: Mix melted beeswax with mineral oil or olive oil and apply it to wooden surfaces for a protective, shiny finish.

Propolis

1. What is Propolis?

Propolis is a resinous substance that bees collect from tree buds and sap flows. They use it to seal cracks, reinforce the hive, and protect it from pathogens.

2. Harvesting Propolis

Scrape Propolis: Collect propolis by scraping it from the hive's frames and inner surfaces using a hive tool.

- **Propolis Traps**: Use propolis traps, which are plastic or metal grids placed in the hive. Bees fill the gaps with propolis, which can then be easily removed.

3. Uses of Propolis

Health Supplements: Propolis is known for its antimicrobial and anti-inflammatory properties.

- **Tinctures and Extracts**: Dissolve propolis in alcohol or glycerin to create tinctures, which can be used for immune support and wound healing.

Topical Applications: Propolis can be used in salves and creams for skin conditions and minor wounds.

- **Recipes**: Mix propolis extract with beeswax and carrier oils to create soothing ointments.

Oral Health: Propolis is used in natural toothpaste and mouthwash formulations for its antibacterial properties.

- **DIY**: Add propolis tincture to homemade toothpaste recipes for added oral health benefits.

Royal Jelly

1. What is Royal Jelly?

Royal jelly is a nutritious secretion produced by worker bees to feed larvae and the queen. It is rich in proteins, vitamins, and amino acids.

2. Harvesting Royal Jelly

Queen Cells: Royal jelly is harvested from queen cells, where it is produced in larger quantities.

- **Step 1: Grafting**: Transfer young larvae into queen cups placed in a queenless hive to encourage the production of queen cells.

- **Step 2: Collecting**: After a few days, remove the queen cells and extract the royal jelly using a small spatula or suction device.

3. Uses of Royal Jelly

Nutritional Supplements: Royal jelly is consumed for its potential health benefits, including boosting energy and supporting the immune system.

- **Capsules and Powders**: Royal jelly is available in capsule or powder form for convenient consumption.
- **Direct Consumption**: Fresh royal jelly can be eaten directly, though it has a strong, slightly bitter taste.

Cosmetic Products: Royal jelly is used in skincare products for its nourishing and rejuvenating properties.

- **Recipes**: Add royal jelly to homemade face creams, masks, and serums to benefit from its anti-aging and moisturizing effects.

Fertility and Hormone Balance: Some people use royal jelly to support fertility and hormonal balance, although scientific evidence is limited.

- **Tinctures and Supplements**: Incorporate royal jelly into daily supplements aimed at promoting reproductive health.

Beekeeping offers a wealth of products beyond honey, each with unique properties and benefits. By harvesting and utilizing beeswax, propolis, and royal jelly, you can fully appreciate the diverse gifts that bees provide. Whether you use these products for health, skincare, or home projects, they add valuable dimensions to your beekeeping experience.

Chapter 5

Pest and Disease Management

Managing pests and diseases is crucial to maintaining healthy bee colonies and ensuring their productivity. This chapter will guide you through identifying common pests and diseases, exploring both natural and chemical treatment options, and implementing Integrated Pest Management (IPM) strategies.

Identifying Common Pests and Diseases

Understanding the common threats to your hive allows you to detect issues early and take appropriate action.

1. Varroa Mites

Description: Small, reddish-brown external parasites that attach to bees and feed on their hemolymph. **Symptoms**: Deformed wings, weakened bees, reduced brood, and visible mites on bees or in brood cells.

2. Small Hive Beetles

Description: Small, black or brown beetles that infest hives, laying eggs that develop into destructive larvae. **Symptoms**: Larvae tunneling through comb, slimy hive interiors, and fermented honey odor.

3. Wax Moths

Description: Moth larvae that feed on beeswax, honey, and pollen, causing damage to combs and hive structures.

Symptoms: Webbing in the comb, tunnels in wooden frames, and destroyed combs.

4. American Foulbrood (AFB)

Description: A bacterial disease caused by *Paenibacillus larvae* affecting bee larvae. **Symptoms**: Sunken, perforated brood caps, a foul odor, and brown, sticky larval remains.

5. European Foulbrood (EFB)

Description: A bacterial disease caused by *Melissococcus plutonius* affecting bee larvae. **Symptoms**: Yellow, twisted larvae, a sour odor, and spotty brood pattern.

6. Nosema

Description: A fungal infection caused by *Nosema apis* or *Nosema ceranae* affecting the bee's digestive system. **Symptoms**: Diarrhea at the hive entrance, reduced colony strength, and slow spring build-up.

7. Chalkbrood

Description: A fungal disease caused by *Ascosphaera apis* affecting bee larvae. **Symptoms**: White, chalky mummies in brood cells, and weakened colonies.

Natural and Chemical Treatment Options

Both natural and chemical treatments can help manage pests and diseases, each with their advantages and considerations.

1. Varroa Mites

Natural Treatments:

- **Drone Brood Removal**: Removing and freezing drone brood frames to reduce mite populations.
- **Screened Bottom Boards**: Using screened bottom boards to allow mites to fall out of the hive.

Chemical Treatments:

- **Formic Acid**: Effective but can be harsh on bees; follow manufacturer instructions carefully.
- **Oxalic Acid**: Administered via vaporization or dribble method; effective during broodless periods.
- **Thymol**: Found in products like Apiguard, it disrupts mite reproduction.

2. Small Hive Beetles

Natural Treatments:

- **Beetle Traps**: Place traps filled with oil or beetle attractants in the hive.
- **Ground Diatomaceous Earth**: Spread around the hive to kill larvae entering the soil.

Chemical Treatments:

- **Permethrin Strips**: Used inside the hive to kill beetles; follow safety guidelines.

3. Wax Moths

Natural Treatments:

- **Strong Colonies**: Maintain strong colonies that can defend against moth infestations.
- **Freezing Equipment**: Freeze combs and equipment to kill larvae and eggs.

Chemical Treatments:

- **Bacillus thuringiensis (Bt)**: A biological control agent sprayed on combs to prevent moth infestation.

4. American Foulbrood (AFB)

Natural Treatments:

- **Hive Destruction**: In severe cases, burn infected hives and equipment to prevent spread.

Chemical Treatments:

- **Antibiotics**: Use oxytetracycline under veterinary supervision to control infections.

5. European Foulbrood (EFB)

Natural Treatments:

- **Requeening**: Introduce a new queen to improve colony health and resistance.

Chemical Treatments:

- **Antibiotics**: Similar to AFB, use oxytetracycline for treatment.

6. Nosema

Natural Treatments:

- **Good Hive Management**: Ensure proper ventilation and reduce stress factors.
- **Fumagillin**: Use where legal and appropriate to control Nosema levels.

7. Chalkbrood

Natural Treatments:

- **Improved Ventilation**: Ensure proper airflow within the hive.
- **Requeening**: Introducing a new queen can help strengthen the colony's resistance.

Integrated Pest Management (IPM)

IPM is a holistic approach combining various methods to manage pests and diseases effectively while minimizing harm to bees and the environment.

1. Monitoring and Identification

- **Regular Inspections**: Conduct frequent hive inspections to detect early signs of pests and diseases.

- **Monitoring Tools**: Use sticky boards, mite counts, and visual checks to monitor pest levels.

2. Cultural Practices

- **Hygiene**: Maintain clean equipment and hives to reduce disease transmission.
- **Strong Colonies**: Focus on keeping strong, healthy colonies that can resist pests and diseases.

3. Mechanical and Physical Controls

- **Screened Bottom Boards**: Use to help reduce mite populations.
- **Bee Traps**: Utilize beetle traps and other mechanical controls to manage pests.

4. Biological Controls

- **Beneficial Organisms**: Introduce organisms like beneficial nematodes to target soil-dwelling pests.
- **Microbial Treatments**: Use biological agents such as Bt for wax moth control.

5. Chemical Controls

- **Targeted Application**: Use chemicals as a last resort and follow integrated guidelines to minimize resistance development.

- **Safe Products**: Choose products that are safe for bees and apply them according to label instructions.

Effective pest and disease management is essential for maintaining healthy, productive hives. By identifying common pests and diseases, utilizing both natural and chemical treatment options, and implementing an Integrated Pest Management (IPM) approach, you can protect your bees and ensure the long-term success of your beekeeping operation.

Hive Management Strategies

Effective hive management strategies are crucial for maintaining healthy colonies and maximizing honey production. This chapter covers essential techniques, including splitting hives, queen rearing basics, and combining weak hives, to help you manage your bee colonies successfully.

Splitting Hives

Splitting hives is a method used to prevent swarming, increase colony numbers, and manage colony size.

1. Reasons to Split Hives

- **Swarm Prevention**: Reduces overcrowding and the natural swarming instinct.

- **Colony Expansion**: Allows you to increase the number of hives.

- **Rejuvenation**: Helps in requeening and breaking the brood cycle to manage pests and diseases.

2. When to Split Hives

- **Spring**: Best time as colonies are growing rapidly.

- **Summer**: Can also be done if there are sufficient resources.

3. How to Split Hives

Step-by-Step Process:

- **Preparation**: Ensure both the parent and the new hive have enough resources (brood, bees, honey, and pollen).

- **Find the Queen**: Determine the location of the queen and decide whether she stays in the parent hive or moves to the new one.

- **Divide Resources**: Move frames of brood, bees, honey, and pollen to the new hive. Ensure both hives have a good mix of these resources.

- **Queen Management**: If you are not moving the queen, ensure the new hive has either a queen cell, a mated queen, or the means to raise a new queen.

- **Location**: Place the new hive near the parent hive, but with a clear distinction to prevent drifting.

4. Post-Split Care

- **Feeding**: Provide supplemental feeding if nectar flow is insufficient.

- **Monitoring**: Regularly inspect both hives to ensure they are healthy and the new queen is accepted and laying eggs.

Queen Rearing Basics

Raising your own queens can improve colony genetics and hive productivity.

1. Why Raise Your Own Queens?

- **Genetic Improvement**: Select for desirable traits such as disease resistance, temperament, and productivity.
- **Cost-Effective**: Reduces the need to purchase queens from external sources.
- **Colony Management**: Ensures availability of queens for splits, requeening, and replacing failing queens.

2. Methods of Queen Rearing

Simple Methods:

- **Swarm Cells**: Utilize naturally occurring swarm cells in strong hives.
- **Supercedure Cells**: Use queen cells from colonies that are naturally requeening.

Advanced Methods:

- **Grafting**: Transfer young larvae into queen cups to be raised in a queenless colony.

- **Cloake Board Method**: A more controlled technique involving a divided hive to raise queens in the upper section.

3. Steps in Queen Rearing

Step-by-Step Process:

- **Starter Colony**: Create a queenless starter colony by removing the queen from a strong hive.
- **Grafting**: Transfer larvae into artificial queen cups and place them in the starter colony.
- **Finisher Colony**: Move the queen cups to a queenright finisher colony once they are accepted and started.
- **Mating Nucs**: Place emerging queen cells in small mating nucs or hives for mating flights.
- **Introduction**: Once mated and laying, introduce new queens to hives needing requeening.

Combining Weak Hives

Combining weak hives can save resources and strengthen colony health.

1. Reasons to Combine Weak Hives

- **Strengthening Colonies**: Creates a single, stronger colony that is more likely to survive.

- **Resource Efficiency**: Makes better use of limited resources and reduces management efforts.

- **Disease Management**: Helps manage colonies showing signs of weakness or disease.

2. When to Combine Hives

- **Fall**: Often combined in fall to ensure strength going into winter.

- **Anytime Weakness is Detected**: Can be done whenever colonies are not thriving.

3. How to Combine Hives

Step-by-Step Process:

- **Inspection**: Ensure neither hive has diseases that could spread.

- **Newspaper Method**: Place a sheet of newspaper between the two hive boxes to be combined. The bees will chew through the paper gradually, reducing aggression and allowing them to merge peacefully.

- **Queen Management**: Decide which queen to keep. If both queens are of similar quality, you may choose one or allow the bees to decide.

- **Placement**: Position the weaker hive above the stronger one using the newspaper method.

4. Post-Combination Care

- **Monitoring**: Regularly inspect the combined hive to ensure the bees are integrating well and the queen is accepted.

- **Feeding**: Provide supplemental feeding if necessary to support the newly combined hive.

Mastering hive management strategies such as splitting hives, queen rearing, and combining weak hives is essential for successful beekeeping. These techniques not only enhance the productivity and health of your colonies but also provide flexibility in managing your beekeeping operation. With careful planning and regular monitoring, you can ensure your bees thrive and your hives flourish.

Maximizing Honey Production

Producing high-quality honey is one of the main goals of beekeeping. This chapter covers essential strategies for maximizing honey production, including understanding nectar flows, supplemental feeding techniques, and best practices for harvesting and processing honey.

Understanding Nectar Flows

Nectar flows are periods when flowers produce abundant nectar, which bees collect and convert into honey. Understanding these flows is crucial for optimizing honey production.

1. **What is a Nectar Flow?**

    - **Definition**: A period when local flowers produce large amounts of nectar.

    - **Significance**: Bees collect and store nectar during these times, converting it into honey.

2. **Factors Influencing Nectar Flows**

Floral Availability: Different plants bloom at different times, providing nectar at various points in the season.

    - **Local Flora**: Familiarize yourself with the blooming cycles of local plants.

    - **Diversity of Plants**: Plant a variety of nectar-producing flowers to ensure continuous nectar flow.

Weather Conditions: Weather significantly impacts nectar production.

    - **Temperature**: Warm temperatures promote nectar production, while extreme heat or cold can inhibit it.

    - **Rainfall**: Adequate rainfall is necessary for plant health and nectar production. Drought conditions can reduce nectar availability.

3. **Maximizing Nectar Flow Benefits**

Timing Hive Expansions: Add supers (honey storage boxes) before the main nectar flow begins to give bees space to store honey.

- **Preparation**: Ensure hives are strong and healthy going into the nectar flow.

- **Regular Inspections**: Monitor hives frequently during nectar flows to manage space and prevent swarming.

Positioning Hives: Place hives near abundant nectar sources.

- **Proximity to Flowers**: Position hives within foraging distance of flowering plants.

- **Diverse Forage**: Encourage a variety of nectar sources by planting diverse flora or positioning hives near diverse landscapes.

Supplemental Feeding Techniques

Supplemental feeding supports bee health and productivity, especially during periods of nectar scarcity.

1. When to Feed

Drought or Scarcity: Provide food during dry spells or when natural nectar sources are scarce.

- **Seasonal Gaps**: Feed during early spring or late fall when natural nectar sources are limited.

- **Colony Stress**: Support weak or newly established colonies with additional feeding.

2. Types of Supplemental Feed

Sugar Syrup: A simple and effective feed for bees.

- **Spring and Summer**: Use a 1:1 sugar-to-water ratio to stimulate brood rearing and colony growth.
- **Fall**: Use a 2:1 sugar-to-water ratio to help bees store food for winter.

Pollen Patties: Provide essential proteins and nutrients.

- **Brood Rearing**: Use during times of low pollen availability to support brood production.
- **Weak Colonies**: Support undernourished or small colonies with added protein.

Fondant or Candy Boards: Solid feeds for winter months.

- **Winter Feeding**: Use fondant or candy boards when liquid feed risks chilling the bees.

3. How to Feed

Feeders: Use appropriate feeders to provide supplemental feed.

- **Entrance Feeders**: Simple and easy to use, but can attract robbers.
- **Top Feeders**: Provide large quantities of syrup and reduce robbing risk.
- **Frame Feeders**: Placed inside the hive, reducing exposure to robbers.

Placement and Timing: Place feeders strategically and monitor feeding.

- **Inside the Hive**: Place feeders within the hive to reduce robbing and exposure.

- **Regular Checks**: Monitor feeders regularly to ensure they are not empty or moldy.

Harvesting and Processing Honey

Proper harvesting and processing techniques ensure high-quality honey production.

1. When to Harvest Honey

Ripeness: Ensure honey is fully ripened and capped by the bees.

- **Capped Cells**: Check that at least 80% of the honey cells are capped.

- **Moisture Content**: Use a refractometer to measure moisture content, ideally below 18%.

Nectar Flow End: Harvest after the main nectar flow to ensure bees have time to ripen the honey.

2. How to Extract and Process Honey

Preparation: Gather necessary equipment and prepare a clean, organized workspace.

- **Uncapping**: Remove the wax cappings from the honeycomb using an uncapping knife or fork.

- **Extraction**: Use a honey extractor (centrifuge) to remove honey from the comb.

- **Filtering**: Strain the extracted honey through a fine mesh or cheesecloth to remove debris and wax particles.

3. Bottling and Storing Honey

Bottling: Transfer the filtered honey into clean, dry jars or containers.

- **Sanitation**: Ensure all equipment and containers are sanitized to prevent contamination.
- **Labeling**: Label jars with the harvest date and honey variety, if applicable.

Storage: Store honey properly to maintain its quality and prevent crystallization.

- **Cool, Dark Place**: Keep honey in a cool, dark location away from direct sunlight and extreme temperatures.
- **Airtight Containers**: Use airtight containers to prevent moisture absorption and preserve flavor.

Maximizing honey production involves understanding nectar flows, implementing effective supplemental feeding techniques, and mastering the art of harvesting and processing honey. By optimizing these aspects of beekeeping, you can ensure a bountiful honey harvest and enjoy the sweet rewards of your efforts.

Chapter 6

Inspecting Your Hive

Regular hive inspections are vital to maintaining healthy and productive bee colonies. Inspections help you monitor colony health, identify problems early, and take necessary actions to support your bees. This chapter covers the essentials of hive inspections, including how to conduct them, what to look for, and common problems and solutions.

How to Conduct a Hive Inspection

1. Preparation

- **Protective Gear**: Wear appropriate beekeeping gear such as a veil, gloves, and a beekeeping suit to protect yourself from stings.

- **Tools**: Gather necessary tools like a hive tool (for prying open the hive and moving frames), a smoker (to calm the bees), and a brush (to gently move bees off frames).

- **Timing**: Choose a warm, calm day for inspections, preferably mid-morning when most foraging bees are out of the hive, making it less crowded and easier to inspect.

2. Initial Steps

- **Approach Calmly**: Approach the hive calmly and slowly to avoid agitating the bees. Bees are sensitive to quick movements and vibrations.

- **Smoking the Hive**: Use a smoker to calm the bees. Light the smoker and puff a few gentle smokes at the entrance of the hive and under the lid. Wait a minute for the smoke to take effect.

3. Opening the Hive

- **Remove the Lid**: Gently pry open the outer cover and inner cover using the hive tool.
- **Check the Top Frames**: Begin by examining the top frames, working your way down through the hive.

4. Inspecting Frames

- **Frame Removal**: Carefully remove one frame at a time, holding it over the hive to avoid dropping bees. Use the hive tool to separate frames that are stuck together with propolis.
- **Examination**: Inspect each frame for brood, honey, pollen, and signs of disease or pests. Hold the frame up to light to see eggs and larvae more clearly.

5. Key Areas to Check

- **Brood Pattern**: Look for a solid brood pattern with eggs, larvae, and capped brood. This indicates a healthy, productive queen.

- **Queen Presence**: Check for the presence of the queen or
- signs of her activity, such as eggs and young larvae. The queen is usually larger and can be found in the brood area.
- **Honey Stores**: Ensure there are adequate honey stores for the bees. Honey should be stored in the upper parts of the frames.
- **Pollen Stores**: Check for sufficient pollen stores, which are essential for brood rearing. Pollen is usually stored around the brood area.
- **Pests and Diseases**: Look for signs of pests like Varroa mites, small hive beetles, and wax moths, as well as diseases like American Foulbrood, European Foulbrood, chalkbrood, and Nosema.

6. Closing the Hive

- **Reassemble Carefully**: Replace frames in the same order and position as you found them.
- **Gently Close**: Close the hive gently, avoiding squashing bees. Replace the inner cover and outer cover.
- **Final Smoke**: Give a final puff of smoke at the entrance to help bees settle down.

What to Look For

1. **Healthy Brood**

    - **Pattern**: A healthy brood pattern is uniform and solid, indicating a strong queen and good colony health.
    - **Eggs and Larvae**: Presence of eggs and young larvae indicates that the queen is active and laying eggs.

2. **Pests and Diseases**

    - **Varroa Mites**: Look for mites on bees and in brood cells. Mites are small, reddish-brown, and attach to bees.
    - **Small Hive Beetles**: Check for adult beetles and larvae in the hive. They are small, dark, and can cause significant damage.
    - **Diseases**: Be vigilant for signs of American or European foulbrood, chalkbrood, and Nosema. Symptoms include discolored larvae, foul smells, and chalky or mummified larvae.

3. **Queen Cells**

    - **Swarm Cells**: Usually found at the bottom of frames, indicating potential swarming.
    - **Supersedure Cells**: Found in the middle of frames, indicating possible queen replacement by the bees.

Common Problems and Solutions

1. Weak Colonies

- **Cause**: Pests, diseases, poor queen performance, or lack of resources.

- **Solution**: Combine with a stronger colony, requeen with a more productive queen, or treat for pests and diseases.

2. Pests and Diseases

- **Varroa Mites**: Implement treatment plans using natural or chemical controls, such as screened bottom boards, drone brood removal, and miticides.

- **Foulbrood**: Treat with appropriate antibiotics under veterinary guidance. Burn severely infected equipment.

- **Chalkbrood**: Improve ventilation and requeen if necessary. Remove infected larvae.

3. Swarming

- **Signs**: Presence of swarm cells, crowded hive, and reduced brood area.

- **Prevention**: Provide additional space by adding supers, split the hive to reduce overcrowding, and manage the queen by clipping her wings or replacing her.

Regular hive inspections are essential for maintaining healthy and productive bee colonies. By carefully examining your hive, you can identify and address issues early, ensuring the well-being of your bees and maximizing honey production. With practice, you'll become proficient at conducting inspections and interpreting what you see, making you a more effective and confident beekeeper.

Beekeeping Seasons and Annual Hive Care

Understanding the seasonal cycles and how they affect your hive is essential for successful beekeeping. This chapter outlines the key activities for each season and provides a comprehensive guide for annual hive care. Each season presents unique challenges and opportunities, and being prepared will help you maintain healthy, productive colonies throughout the year.

Spring: Growth and Preparation

1. Early Spring

- **Inspection**: Perform a thorough inspection to assess colony health and stores. Check for the presence of the queen, brood pattern, and sufficient food stores.

- **Feeding**: Provide supplemental feeding if necessary, using sugar syrup (1:1 ratio) and pollen patties to stimulate brood rearing and support the bees until natural nectar and pollen are available.

- **Disease Check**: Check for signs of diseases and pests, treating as needed. Focus on Varroa mite levels and other common issues.

2. Mid to Late Spring

- **Hive Expansion**: Add supers (additional boxes) to give bees space for nectar storage and to prevent overcrowding and swarming.

- **Swarm Prevention**: Monitor for swarm cells and manage hive population through splits or other methods. Swarming typically occurs when the hive becomes too crowded.

- **Queen Management**: Assess queen performance and replace her if necessary. A strong, productive queen is essential for colony health.

Summer: Honey Production

1. Nectar Flow

- **Supering**: Ensure adequate super space for honey storage. Add more supers as needed to keep up with the bees' nectar collection.

- **Inspection**: Continue regular inspections, focusing on honey production and pest management. Check for signs of swarming, queen health, and overall hive activity.

- **Ventilation**: Ensure proper hive ventilation to reduce moisture and maintain hive health. Use screened bottom boards and adjust hive entrances if needed.

2. Late Summer

- **Harvesting**: Begin honey harvesting when honey is capped and moisture content is appropriate (below 18%). Use a refractometer to measure moisture levels.

- **Feeding**: If nectar flow is low, provide supplemental feeding to support the bees. Use sugar syrup (1:1 ratio) if necessary.

Fall: Preparation for Winter

1. Early Fall

- **Inspection**: Perform a thorough fall inspection, checking for queen health, brood pattern, and honey stores. Ensure the colony is strong and healthy going into winter.

- **Disease Management**: Treat for pests and diseases, focusing on Varroa mite control. This is a critical time to ensure the colony is healthy before winter.

2. Late Fall

- **Feeding**: Ensure bees have sufficient stores for winter. Provide sugar syrup (2:1 ratio) or fondant as needed to help them build up their reserves.

- **Reduce Hive Size**: Remove excess supers and reduce the hive size to help bees maintain warmth. Combine weak colonies if necessary to ensure they have enough bees to generate heat.

Winter: Maintenance and Monitoring

1. **Insulation**

    - **Hive Insulation**: Insulate hives to protect against extreme cold and wind. Use insulation wraps or materials designed for beekeeping.

    - **Ventilation**: Ensure proper ventilation to prevent moisture buildup. Use upper entrances or ventilation holes to allow moist air to escape.

2. **Monitoring**

    - **Periodic Checks**: Monitor hives periodically for signs of moisture, starvation, and hive health. Use a stethoscope or thermal camera to check for bee activity without opening the hive.

    - **Feeding**: Provide emergency feeding if necessary, using fondant or candy boards. Avoid opening the hive frequently, which can disturb the bees and cause heat loss.

3. **Minimal Disturbance**

    - **Avoid Inspections**: Minimize hive disturbance to help bees maintain warmth. Only open the hive if absolutely necessary, such as for emergency feeding.

Annual Hive Care Summary

- **Spring**: Focus on growth and preparation. Perform inspections, provide supplemental feeding, expand the hive, and prevent swarming.
- **Summer**: Optimize honey production. Add supers, monitor hive health, and harvest honey.
- **Fall**: Prepare for winter. Inspect hives, manage diseases and pests, and ensure sufficient food stores.
- **Winter**: Maintain and monitor. Insulate hives, ensure ventilation, and provide emergency feeding if needed.

Successful beekeeping requires adapting to the seasonal needs of your hives and performing regular inspections to ensure hive health. By understanding the specific tasks for each season and diligently monitoring and caring for your bees, you can maintain healthy, productive colonies throughout the year. Each season presents unique challenges and opportunities, and being prepared will help you navigate them effectively.

Conclusion

As we reach the end of this comprehensive guide, "Beekeeping for Beginners," it's essential to reflect on the journey you've embarked upon. Beekeeping is more than a hobby; it's a profound commitment to understanding and nurturing one of nature's most fascinating creatures. Whether you're drawn to the promise of harvesting your own honey, the environmental benefits, or the simple joy of working with bees, this journey offers countless rewards.

The Deep Connection with Nature

Beekeeping fosters a deep connection with nature. From the moment you first open your hive, you enter a world that is both intricate and awe-inspiring. The rhythmic hum of bees, the sweet scent of honey, and the sight of bees diligently tending to their hive are experiences that bring you closer to the natural world.

This connection isn't just emotional; it's also educational. Beekeepers learn to observe weather patterns, plant cycles, and the subtle shifts in their environment that affect their hives. This heightened awareness enhances your appreciation for the delicate balance of ecosystems and the critical role bees play in pollination and biodiversity.

A Lesson in Patience and Persistence

Beekeeping is a practice that demands patience and persistence. It's not about instant gratification but about long-term commitment and continuous learning. Each season brings new challenges and opportunities, teaching you to adapt and grow alongside your bees.

Your first year may be filled with uncertainty as you learn the basics of hive management, pest control, and honey harvesting. Mistakes are inevitable, but each one is a learning opportunity. Over time, you'll develop a deeper understanding of bee behavior, seasonal cycles, and the nuances of hive health. This knowledge is cumulative, building year after year, making you a more skilled and confident beekeeper.

The Joy of Honey Harvesting

One of the most rewarding aspects of beekeeping is harvesting honey. The first time you extract honey from your hive, it's a moment of triumph and fulfillment. The golden, viscous liquid represents months of hard work by your bees and diligent care from you. Each jar of honey is a testament to the symbiotic relationship between you and your bees.

But honey isn't the only treasure. Beeswax, propolis, and royal jelly are valuable products that can be harvested from your hive. These byproducts can be used to make candles, balms, and health supplements, adding another layer of fulfillment to your beekeeping endeavors.

Environmental Stewardship

By keeping bees, you're contributing to environmental stewardship. Bees are essential pollinators, and your efforts help support local ecosystems and agriculture. Healthy bee populations ensure the pollination of fruits, vegetables, and wild plants, promoting biodiversity and food security.

Moreover, beekeeping encourages sustainable practices. You'll find yourself advocating for pesticide-free gardening, planting bee-friendly flowers, and educating others about the importance of bees. Your beekeeping practice becomes a catalyst for broader environmental awareness and action within your community.

Building Community Connections

Beekeeping also opens doors to new communities. Local beekeeping associations, online forums, and social media groups are excellent resources for sharing knowledge, seeking advice, and finding support. These connections can be invaluable, especially when you're facing challenges or celebrating milestones.

Mentorship is another significant aspect. Experienced beekeepers often take beginners under their wing, offering guidance and encouragement. As you gain experience, you too may find joy in mentoring new beekeepers, passing on the knowledge and passion that you've cultivated.

Challenges and Solutions

Beekeeping is not without its challenges. Pests, diseases, and adverse weather conditions can threaten your hives. However, each challenge is an opportunity to deepen your understanding and improve your practices. Regular hive inspections, effective pest management, and staying informed about beekeeping best practices are essential strategies for overcoming these obstacles.

Adaptability is key. Bees are resilient creatures, but they need a beekeeper who can respond to changing conditions and provide the necessary support. Whether it's adjusting feeding practices, implementing new treatments, or modifying hive structures, your ability to adapt will ensure the health and productivity of your colonies.

Continuous Learning and Innovation

The field of beekeeping is constantly evolving, with new research and innovations emerging regularly. Staying informed through books, scientific journals, and beekeeping conferences will keep you abreast of the latest developments. Embrace a mindset of continuous learning, and be open to experimenting with new techniques and technologies.

Innovation in beekeeping can lead to more efficient and sustainable practices. Whether it's adopting organic beekeeping methods, using advanced hive monitoring systems, or exploring new hive designs, being at the forefront of beekeeping innovation can enhance your practice and benefit your bees.

The Future of Beekeeping

As you look to the future, consider how your beekeeping practice can grow and evolve. Perhaps you'll expand your apiary, start a beekeeping business, or engage in bee conservation efforts. The skills and knowledge you've gained can be applied in various ways, each contributing to the health and sustainability of bee populations.

Beekeeping also offers opportunities for advocacy and education. Sharing your experiences with others, whether through local workshops, school programs, or online platforms, can inspire more people to take up beekeeping or support bee conservation initiatives. Your passion and knowledge can have a ripple effect, promoting greater awareness and action for bees and the environment.

Final Thoughts

Beekeeping is a journey of discovery, growth, and fulfillment. It's a practice that challenges you, rewards you, and connects you to the natural world in profound ways. As you continue your beekeeping journey, remember that each hive inspection, honey harvest, and seasonal transition is part of a larger story of care, stewardship, and partnership with one of nature's most remarkable creatures.

Your commitment to beekeeping not only benefits your hives but also contributes to the well-being of our planet. By nurturing your bees, you are playing a vital role in supporting ecosystems, promoting biodiversity, and ensuring the future of pollination.

Thank you for embarking on this journey with "Beekeeping for Beginners." May your beekeeping endeavors be filled with joy, learning, and the sweet rewards of honey. Happy beekeeping!

www.ingramcontent.com/pod-product-compliance
Lightning Source LLC
Chambersburg PA
CBHW052331220526
45472CB00001B/379